Alexander Fleming and the Story of Penicillin

John Bankston

PO Box 619
Bear, Delaware 19701

Unlocking the Secrets of Science

Profiling 20th Century Achievers in Science, Medicine, and Technology

Alexander Fleming and the Story of Penicillin

Mitchell Lane PUBLISHERS

Copyright © 2002 by Mitchell Lane Publishers, Inc. All rights reserved. No part of this book may be reproduced without written permission from the publisher. Printed and bound in the United States of America.

Printing   3   4   5   6   7   8   9   10

Library of Congress Cataloging-in-Publication Data

Bankston, John, 1974-

      Alexander Fleming and the story of penicillin/John Bankston.

          p. cm. —(Unlocking the secrets of science)

      Includes bibliographical references and index.

      Summary: A biography of the Scottish bacteriologist and Nobel Prize winner who discovered penicillin and its antibiotic properties.

      ISBN 1-58415-106-4

      1. Fleming, Alexander, 1881-1955—Juvenile literature. 2. Bacteriologists—Great Britain—Biography—Juvenile literature. 3. Penicillin—History—Juvenile literature. [1. Fleming, Alexander, 1881-1955. 2. Scientists. 3. Penicillin—History. 4. Nobel Prizes—Biography.] I. Title. II. Series.

QR31.F5 B36 2001

616'.014'092—dc21

[B]   2001042772

ABOUT THE AUTHOR: Born in Boston, Massachusetts, John Bankston began publishing articles in newspapers and magazines while still a teenager. Since then, he has written over two hundred articles, and contributed chapters to books such as *Crimes of Passion* and *Death Row 2000*, which have been sold in bookstores around the world. He has recently written a number of biographies for Mitchell Lane including books on Mandy Moore, Jessica Simpson and Jonas Salk. He currently lives in Los Angeles, California, pursuing a career in the entertainment industry. He has worked as a writer for the movies Dot-Com and the upcoming *Planetary Suicide*, which begins filming in 2002. As an actor John has appeared in episodes of *Sabrina the Teenage Witch*, *Charmed* and *Get Real* along with appearances in the films *Boys and Girls*, and *America So Beautiful*. He has a supporting part in *Planetary Suicide* and has recently completed his first young adult novel, *18 To Look Younger*.

PHOTO CREDITS: cover: Archive Photos; p. 6 Hulton/Archive; p. 10 Library of Congress/Science Source; p. 16 Archive Photos; p. 24 St. Mary's Hospital Medical School; p. 31 Archive Photos; p. 46 Library of Congress; p. 52 Globe Photos.

PUBLISHER'S NOTE: In selecting those persons to be profiled in this series, we first attempted to identify the most notable accomplishments of the 20th century in science, medicine, and technology. When we were done, we noted a serious deficiency in the inclusion of women. For the greater part of the 20th century science, medicine, and technology were male-dominated fields. In many cases, the contributions of women went unrecognized. Women have tried for years to be included in these areas, and in many cases, women worked side by side with men who took credit for their ideas and discoveries. Even as we move forward into the 21st century, we find women still sadly underrepresented. It is not an oversight, therefore, that we profiled mostly male achievers. Information simply does not exist to include a fair selection of women.

Contents

Never one to worry about neatness, Alexander Fleming often grew cultures in unwashed petri dishes, such as the one he is holding in this photo.

Chapter 1

A Lucky Mess

• •

Have you ever gotten in trouble for not picking up your room? The next time that your parents tell you to clean up your room, you can tell them that a messy room was responsible for the discovery of penicillin, one of the true miracle drugs of the 20th century.

Before penicillin's discovery in 1928, soldiers on the battlefield were more likely to die from infection than from bullets. Surgeons would perform life-saving operations, only to watch helplessly as their patients died from infections soon afterward.

And it wasn't just wounded soldiers who were at risk. Diseases such as tuberculosis, cholera and diphtheria killed tens of thousands of people.

In the 1800s, French scientist Louis Pasteur was the first to theorize that specific diseases and infections were caused by specific germs—microscopic single-celled organisms called bacteria.

In the 1870s and 1880s, the work of German scientist Robert Koch led to a new branch of medicine. Known as bacteriology, it is the study of the more than 2,500 different types of these little creatures. They are so small over a billion could fit in a teaspoon of dirt. Most of them are harmless, even helpful. However, a few are very dangerous.

A bacteria that is both helpful and harmful is *E. coli*, which was in the news in 2001. Some people became sick and a few even died—most of them children—after eating fast-food hamburgers which were contaminated with *E. coli.*

Normally, *E. coli* is a bacteria that exists in our intestines and helps to produce some essential vitamins from the food we eat. But a strain, or variation, of *E. coli* recently developed that produced a toxin which attacked intestinal walls and caused severe internal bleeding. Small children were especially at risk because of their relatively limited blood supply.

Another harmful bacteria is known as *Streptococcus pyogenes.* This bacteria grows in clumps, like a bunch of grapes. Like all bacteria, it reproduces by cell division— splitting in half so one cell becomes two, two become four, four become eight, and so on. Since bacterial cell division can occur every twenty minutes, the organism spreads quickly. Until the 1900s, infection by the streptococcus was often deadly. Even the tiniest infected cut could be fatal.

One of the diseases it causes is strep throat, a throat infection that commonly occurs in children. It is highly contagious, being spread through airborne droplets when someone with the disease coughs or sneezes. It can also be picked up by touching infected surfaces.

Strep throat is not only painful but can also lead to rheumatic fever, which in turn can result in serious heart or joint damage. A rare complication of rheumatic fever can cause uncontrolled movement of the arms and legs which is known as "St. Vitus dance."

One reason why battlefields were so dangerous was the *Clostridium welchii* bacteria. It produces spores which are like seeds and can survive for years, and is responsible for a condition known as gas gangrene.

During wartime, shrapnel—sharp pieces of an exploding bomb or other artillery—not only injures soldiers, but the hot metal fragments often drive bits of dirty clothing, ground debris and even vegetation which contain the *Clostridium welchii* bacteria into their wounds. The resulting gas gangrene infection would then spread along the muscle, turning the affected area black and expanding it with gas. Before the 1920s, the only way to save a life of a soldier with a gangrenous wound was through amputation—cutting off the infected arm or leg.

Often the infection was discovered too late. Even after the harsh remedy of losing a limb, a wounded solder would still die.

Today, the world is much safer. While some people still die from bacterial infections, the percentage is only a very tiny fraction of what it was less than a century ago.

And one of the main reasons is that a 47-year-old research doctor didn't clean up his London laboratory and left a window wide open when he took a two-week vacation in the late summer of 1928.

His name is Alexander Fleming, and this is his story.

Alexander Fleming endured the disbelief of his peers and would not be recognized for his discoveries until he was well into his sixties.

Chapter 2

A Scottish Childhood

• •

Ayrshire County in southwestern Scotland is an area of fertile coastal land and barren hills. Along the valleys, winter winds blow up to one hundred miles per hour. Year round, thick fog covers the land and rain falls with almost monotonous regularity. It is a place where the harsh challenges of the land greatly influence the lives of those who live there. Even today, farming is often difficult because of the weather and the unique geography.

But despite these difficulties, the Fleming family—who took their name from the Flemish people of Belgium where they came from—had lived in Ayrshire for several centuries. The family had two skills: weaving and farming.

Like his father before him, Hugh Fleming was a farmer. Hugh was in his sixties when he remarried; Jane, his first wife, had died from tuberculosis several years earlier.

On August 6, 1881, Grace Fleming—Hugh's second wife—gave birth to a son, Alexander Fleming. Alec, as he would soon be known, was born inside his family's isolated farmhouse four miles from Darvel, the nearest town.

He was the seventh of eight children. In addition to Alec there were three other boys and three girls: Hugh III, Thomas, and John, along with Jane (who as a young woman died of smallpox), Mary and Grace. Alec was the next to last child—his younger brother Robert would be born in 1883.

Sadly, all Alec would remember of his father was a gentle old man who spent most of his time confined to a chair by the fire. He died from a stroke when Alec was seven. As the eldest son, Hugh III took over the farm.

The Flemings' farm had over three hundred sheep to care for and shear in season, along with dairy cows to be milked, chickens to be raised and other chores. Alec grew up in a house lit by candles and paraffin lamps, where there was no indoor plumbing and bathing was done in a tin tub in the kitchen, the water heated on a wood stove.

Despite these hardships, Alec enjoyed his life and the close contact with nature that wandering the farm's eight hundred acres of rough land provided. Although Alec got along well with others, he enjoyed his own company, spending many hours by himself exploring and studying. In a biography written about him, he would say, "As boys we had many advantages over the boys living in towns. We automatically learned many things that they missed, and it was just the chance of living away from people which taught us those things."

Although farm life was often difficult, there was still plenty of time for evening parlor games. The Flemings would often invite their closest neighbors, the Loudons, over for these games and festive parties during Christmas and Halloween.

At the age of five, Alec began attending a "wee" school, a one-room schoolhouse in Loudon Moor. Alec would say he spent "the brightest part of my schooldays" in the tiny room, kept warm with the thick strips of peat which grew near the children's homes. There were only ten or so students, as his classmates were mainly other Flemings and the Loudons.

Alec was small for his age, but athletic, a tow-headed youngster best remembered for both his penetrating blue eyes and quiet intelligence. One of his teachers, Marion Stirling, was so impressed that she told friends "young Fleming will get somewhere before he's many years older."

At ten, Alec began attending a larger school in Darvel. Like many other children from Loudon Moor, he quickly advanced past the level of his new classmates. Although it was tiny, his old school had developed a reputation for preparing its students very well.

The new school wasn't just bigger, it was also four miles away. Fleming's only way to get there was to walk. On many mornings he'd leave his warm farmhouse in the dark. Even in the spring, those early morning walks were chilly, but his mother Grace cooked up a solution. When little Alec left for school, she'd hand him a pair of baked potatoes fresh from the oven and he'd put them in his coat pockets. Not only did they keep him toasty on his long trek to school, they also provided him with lunch at noon!

Alec began attending middle school at Kilmarnock Academy when he was twelve. It was too far away for him to walk, so he'd board with his aunt during the week and take a train home every Friday.

Meanwhile, Alec's older brothers Tom and John moved from the farm in Scotland to the huge city of London. It was a powerful, cosmopolitan place in the early 1890s—a world leader in arts, culture and business. Soon Alec's oldest brother Hugh decided that it was time for Alec to head for "the big city" as well.

So at fourteen, Alec left Scotland and moved in with Tom, who had just gone into business as an oculist, or eye doctor, and owned a house on York Street. John was there as well, learning the lens trade from an eyeglass manufacturer. The young men were cared for mainly by their sister Mary, who ran the household.

It was a good homey environment for Alec, who enrolled in the prestigious Regent Street Polytechnic Institute. Six months later, he was joined by his younger brother Robert. The high educational standards of Scotland helped Alec to quickly move up four classes at Polytechnic.

He also quickly developed a reputation for very casual studying habits. While his classmates would pore over their textbooks for hours, Alec would spend his evenings playing games with the family. Late at night or just before school in the morning, he would briefly thumb through his work. Despite this relaxed attitude, he got high scores on all of his exams.

Unfortunately, the cost of his education was an enormous burden to his older brother Tom, who was still trying to build up his practice as an eye doctor. Money was tight and Alec felt guilty.

At age sixteen, the young man who many teachers thought was destined for a bright future left school. To help out his family, Alec took a position as a clerk in a shipping firm. It was a job with limited opportunities and almost no use for his intellect.

The teenager with so much promise had become a dropout. In 1897 it looked like Alec Fleming's life would be filled by little more than menial work and daydreams.

While many twentieth century scientists devoted most of their early years to academics, young Alexander Fleming often preferred sports to studying.

In the 1790s, Edward Jenner discovered the power of vaccines by injecting healthy children with cow pox virus to prevent smallpox. Fortunately, the enormous risk paid off.

Chapter Three
A Second Chance

· ·

S tuck in a shipping office, filling out cargo lists as a junior clerk for fifteen shillings a week, Alec Fleming's days were filled with teeth-numbing boredom. Although his work life was tedious, Alec's free time was considerably more interesting. He joined the London Scottish Volunteers. Like the National Guard in the United States, the volunteers were part-time soldiers with regular jobs. In addition to military training, these weekend warriors donned kilts—skirt-like traditional Scottish garments—and played bagpipes, their homeland's most famous musical instrument.

While a member of the Volunteers, Alec took up sports, playing water polo and engaging in shooting competitions. He won a number of trophies, and would later tell a biographer that "There is no better way to learn about human nature than by indulging in sports, more especially team sports."

Fleming's love for sports wouldn't just teach him about human nature. It would also change his life on several occasions.

Four years after he dropped out of the Polytechnic Institute, Fleming received a financial boost when his Uncle John died and left Alec over two hundred pounds (the British pound is worth about one and a half times the United States dollar) in his will. A century ago, two hundred pounds was a great deal of money. Alec realized he had the chance to leave behind the drudgery of the shipping office. But he was clueless about what to do with his life. It was his brother Tom—by now a successful eye doctor—who suggested that Alec give medicine a try.

Alec wasn't sure he wanted to be a doctor. He just knew he didn't ever want to fill out another cargo slip. But getting into medical school required passing a very difficult entrance exam and Alec hadn't taken a test in four years. Despite this challenge, Alec again barely studied, relying on a few quick study courses in medicine.

Alec easily passed the exam and his score was so high that he had his pick of schools. But he didn't know much about any of them. What he did know was that the water polo team at St. Mary's—one of the schools he was considering—had defeated the team he played for. So Alec based his decision on sports. Because St. Mary's had a good water polo team, that was where Alec would study to be a doctor.

St. Mary's Hospital Medical School was founded in 1845 and was the newest of London's twelve teaching hospitals. Unfortunately, its reputation in water polo was about the only thing it had going for it. Burdened by large debts, the school was always one step away from being shut down. Inside its building on South Wharf Road, the hallways and classrooms were dark and dirty.

The hospital and its school next door were located in the Paddington section of London. A once-fine little village, Paddington was by the late 1800s an industrial slum, often called "stinking Paddington" by other Londoners.

The area was filled with decaying homes and the shacks of railroad and dock workers. These homes lacked plumbing or sanitation, and most residents would start their mornings by dumping buckets of waste water into the streets. Because of this, it was an area where bacterial infections flourished.

It might have been a terrible neighborhood. But it was the perfect place for a future bacteriologist to begin his medical education.

In 1901, Alec joined a class of 79 other medical students. Most of them were younger than he was, and had come right from high school. Still, Alec knew his life was changing for the better. His fellow students might have school experience, but he had *life* experience.

Just as important, Alec had a gift. Throughout his academic career, he was able to absorb the required material with half the effort that other students put in. He instinctively grasped what the professors and the books he was reading were trying to say. Sometimes others would watch as Alec casually thumbed through a book and didn't even look as if he was reading it, except by his occasional grimace and a comment about how wrong the author was.

A fellow student, Charles Pannett, would later tell Fleming's biographer, "He never burdened himself with unnecessary work, but would pick out from his textbooks just what he needed and neglect the rest."

Alec didn't just pass his courses, he won awards for his academic achievement. Besides the anatomy prize and the physiology prize, Alec was also awarded an entrance scholarship which paid 150 pounds and helped to cover his medical school expenses.

It wasn't just his ability to easily absorb important information which set him apart from his classmates. The life of most medical students a century ago was very different from today. Students could take as long as they liked to complete the required courses and exams. Some took eight, even ten years to finish. With so much free time, medical students were well-known for their drunken, rowdy behavior.

Alec wasn't like that. Perhaps it was being a farmer's son that set him apart from more privileged classmates. Maybe it was his nature. In some ways, Alec stood out by not standing out. Standing just five feet, five inches as an adult, he was shorter than his peers and was soon nicknamed "Little Flem."

In later life, a colleague told Fleming that his son was studying hard for an exam, Fleming replied, "He doesn't need to bother about exams—he's tall. Tall people can do anything, go anywhere." Not just shorter than most of his peers, he was also quieter, the kind of person who would let others go on and on, his intense blue-eyed stare never wavering.

When Alec did speak, he made it count. Alec's talent for saying the right thing would serve him well when a controversial doctor named Almroth Wright joined the faculty at St. Mary's. In Alec's second year of medical school, Wright became a professor of pathology, which is the study of disease.

Four years before, Wright became the first doctor to use a killed vaccine to inoculate people—especially soldiers—to fight typhoid fever. Unfortunately, his efforts to get the vaccine accepted by the Army were so unsuccessful he eventually resigned in disgust.

In the early part of the twentieth century, Almroth Wright was at the forefront of research in inoculation. When people are inoculated by weakened or dead germs, they produce what are known as antibodies—natural defenses against these germs. A good example is when a doctor gives a patient a shot for measles. The person doesn't get measles from the shot and afterwards is protected from the disease for a period of time.

The process had been discovered over a century earlier. In 1796, Britain's Edward Jenner noticed milkmaids coming into contact with "cowpox" from sores on a cow's udder. After this contact, the milkmaids didn't get smallpox. Jenner scraped off some of the cowpox sores and injected it into an eight-year-old boy. As expected, the boy got sick, but he recovered. Six weeks later, the doctor injected the same boy with smallpox sores. The boy stayed healthy—he was now *immune* to smallpox.

In 1798, Jenner injected hundreds of people with cowpox during a devastating smallpox epidemic. Like the boy, they all became immune to the disease. In a book he wrote on the subject, Jenner called the technique "vaccination." The word is derived from the Latin word "vacca," for cow. Although vaccinations no longer involve cows, the term is still used.

Wright believed in the power of vaccination and valued it over all other medical discoveries. He believed a doctor's greatest role in the future would be as an immunizer—preparing vaccines and giving people shots.

"Of all the evils that befall man in his civilized state, the evil of disease is incomparably the greatest," Wright wrote in the *Liverpool Daily Post.* "If the belief is nurtured that the medical art of today can effectively intervene in the course of disease, this ought to be dismissed as illusion."

Wright's conviction that doctors were ripping off their patients by offering cures that didn't work made other doctors very angry and he was unpopular among them.

With a marred reputation, the man called "Almost Wright" by his critics was perfect for a school whose claim to fame was a world class polo team. Joining St. Mary's was a fresh start for the forty-year-old Wright.

Fleming and Wright didn't get together immediately. Wright was focused on bacteriology and vaccination, while Fleming loved surgery. Surgeons were the doctors with the most respect and the highest pay. With his abilities as a quick study and the type of hand-eye coordination he demonstrated in water polo and riflery, Fleming was a natural for surgery. Besides, research was considered beneath most medical students, who did it grudgingly in school and rarely as professionals.

In 1905, Fleming passed the Primary Examination for the Royal College of Surgeons. He planned to focus on delivering babies, and spent a month going to the crowded homes of Paddington's poorest residents, often delivering an infant in a tiny room surrounded by a dozen relations.

But he also spent time in the hospital, dealing with the types of infections caused by bacteria. These bacteria were responsible for deadly diseases like tuberculosis—which had killed his father's first wife—and lobar pneumonia, which flourished in "stinking Paddington."

As a result, Fleming began attending many of Wright's lectures; the subject of bacteriology fascinated young Alec. Still, when he graduated, he planned to pursue a career in surgery.

But for the second time, athletic competition would once again change the direction of Alec Fleming's life.

St. Mary's was trying to win the Armitage Cup, a trophy given to the winner in rifle competition, and Fleming was one of the team's top shooters. If he graduated and left St. Mary's, the team would lose one of its best men. St. Mary's wanted to win the trophy, so strings were pulled. Fleming was offered a chance to be a junior assistant in Almroth Wright's laboratory.

Because he wanted to help the rifle team win a trophy, Fleming took a job at the lab where two decades later he would make the discovery that would change medicine and save millions of lives.

Although older than many of his fellow students, Alexander Fleming was able to endure long hours and primitive conditions because of his childhood spent on a farm.

Chapter Four

Private 606

● ●

Joining the inoculation department run by Almroth Wright meant two things for 25-year-old Alexander Fleming: very long hours and being surrounded by a great, swirling hurricane of controversy.

The long hours came from Wright's enormous amount of energy and passion for his subject. He had what he called a "pain in the mind" because of his overwhelming drive to help others through his research. Fleming was mainly there because St. Mary's needed a skilled marksman, but Wright's enthusiasm was as contagious as tuberculosis.

Fleming didn't just toil until midnight preparing slides and studying microbes under the microscope. Like the other assistants, he was expected to join Wright at the end of the evening. There Wright would spend hours leading discussions on everything from poetry to science. Although these sessions could go on until 2:00 or 3:00 in the morning, the research assistants were all expected to be back at the lab by 9:00 a.m. sharp.

Fleming had an advantage over his peers. He'd already developed his stamina during those long walks when he was a child. Further, while the others struggled to seem intelligent, Fleming stayed quiet, waiting for the few times when Wright asked his opinion.

In the beginning Wright didn't appreciate the shy and quiet Fleming. The older doctor often said, "There are three degrees of discourtesy in conversation: to contradict, to change the subject and to receive it in silence."

Fleming overcame his shyness once he realized most of the poetry Wright quoted was from a single author. After that, whenever Wright would recite a verse or two, Fleming would confidently name the same author. Most of the time, he was right.

One of Fleming's first jobs in the lab was studying white blood cells. These are cells that defend the body against invading bacteria and other germs that cause infections.

In 1882, the Russian-French bacteriologist Elie Metchnikoff gave the particular white blood cells that eat invading bacteria the name of "phagocytes." When Fleming joined Wright's staff, the doctor was busy conducting research into the various properties of phagocytes and the process of phagocytosis, which occurs when phagocytes consume invading microbes.

Wright also discovered serum, the clear liquid part of the blood, that made the microbes "tasty." He called it "opsonin," which is Greek for "I prepare food." He developed an "opsonic index," a method of determining how much of the serum was present in blood. Wright believed high amounts of this serum would increase the effectiveness of a vaccine.

As one of his first duties at the lab, Fleming had to do the very tedious job of determining the opsonic index in numerous blood samples. Although somewhat boring, it was important work. Learning how our bodies naturally defend themselves against bacteria was an important first step in developing the means to protect people from illness.

As an assistant in Wright's inoculation lab, Fleming was well aware of the controversy that surrounded his boss.

However, Fleming managed to avoid most of it by his quiet, thoughtful manner. He was popular with the others because like many quiet people, he was considered a good listener, although his habit of staring directly at the speaker and then leaving without a word once the conversation ended was a bit unsettling. Still, Fleming was well-liked.

Wright wasn't.

By going against the standards of the time, Wright attracted many critics. Not surprisingly, he was embraced by those whose careers often rely on going against those standards—writers and artists. Wright had many friends in the arts, including the well known playwright George Bernard Shaw. Wright was even the basis for the lead character in Shaw's play *The Doctor's Dilemma*.

Fleming was also popular with artists. When painter Ronald Grey became ill, Fleming treated him. The two became so friendly that Grey eventually nominated Fleming for membership in the Chelsea Arts Club. Although Grey wanted Fleming to be the club's "Honorary Bacteriologist" and treat other members for free, the club required that those who joined earn at least some of their income as artists.

To satisfy the requirement, Fleming painted an abstract art piece. He felt that the odd painting was a subtle comment on the "modern art" movement, where many paintings looked very little like their subjects. To his surprise, one newspaper complimented the work's "sophisticated naivety." Unfortunately the painting didn't sell and eventually Fleming loaned money to a friend to buy the work.

Still, Fleming would continue to develop his skills as an artist. He even turned his talent for glass blowing—which he used in the lab to make his own test tubes and culture plates—into art. When a pair of shy St. Mary's students came to learn the technique, Fleming took the time to show them the method. Then he surprised them when instead of producing lab equipment, he crafted a tiny blown glass sculpture of a cat chasing several tiny mice.

Fleming also began using germ cultures as the basis for later paintings. To Fleming this combination of art and science made perfect sense. "Most scientists are artists in a sense," he later wrote. "Unless they have vision they can do comparatively little with their formula."

It was Fleming's vision that continued to move his career forward. While he continued to perform research under Wright's supervision, he branched out into other areas that interested him. He began to focus on a condition that concerns many teenagers: acne. By studying the bacteriology of pimples, Fleming discovered the *Corynebacterium acnes* bacteria. He made a culture—a growth designed especially for study—from this bacteria. Then he used three strains (varieties) of the bacteria and produced a vaccine that temporarily cleared up the condition. In 1909 he would publish his findings in the famous British medical journal, *The Lancet*.

In June of that year, he took the final examination for the Royal College of Surgeons. Fleming passed, and was qualified to take a surgical staff position. Although the job would be both prestigious and high-paying, he decided to stay on as a research assistant. Later, when someone asked why he even bothered to take the exam if he never intended to be a surgeon, Fleming is quoted as saying, "I never ceased to regret the five pounds which I spent on no purpose. I wondered whether I ought not, perhaps, to give a shot at the final."

While he never did another operation once he was fully qualified to do so, Fleming made huge strides in his research work. Besides his work on the acne vaccine, in 1910 Fleming was offered the chance to help treat a far more serious condition: syphilis.

Syphilis is a disease transmitted by sexual intercourse or from an infected mother to her unborn baby. Caused by a spirochete—a wavy, very mobile bacteria—syphilis was once called "the pox." Untreated, an infected person would suffer through three distinct phases of the disease, each more horrible than the last. The first phase was a small sore, while the second phase a few months later was widespread skin ulcers. Following that came the third phase—slow destruction of the nerves, blood vessels, bones and even the brain. It was usually a horrible, incurable, fatal disease.

All of that changed in the early 1900s.

In 1909, Fleming was part of the team that improved the very difficult methods of diagnosing syphilis in a laboratory, before most of the unpleasant symptoms of the disease appeared. He developed a method of diagnosing the disease using only a few drops of blood and published three papers detailing his work. His skill at diagnosing the condition made him much in demand by both London hospitals and private physicians.

In 1910, a German doctor named Paul Ehrlich completed his work creating a vaccine which would kill the syphilis microbes but not harm the patients. He tested 605 compounds before he found the one which worked. Its scientific name was dioxy-diamino-arsenobenzene, which Ehrlich nicknamed "606."

Wright was a good friend of Ehrlich, and his lab was one of the first chosen to do clinical trials of the new vaccine. Alec Fleming, water polo champion and crack shot, had a great deal of skill with his hands. Already recognized for his diagnostic abilities, Fleming quickly developed a reputation for his ability with a needle. Because his boss was one of the few people with a steady supply of 606, and because Fleming was so skilled, he built a private practice inoculating hundreds of syphilitic patients.

Fleming's artist friend Ronald Grey even published a cartoon of the young doctor dressed in the traditional kilts of his London Scottish Volunteers, holding a large needle instead of a rifle. The cartoon was labeled "Private 606," a reference not only to the drug he was administering, but also to his military rank. Despite well over a decade of service, Fleming still hadn't advanced beyond private. All of that was about to change.

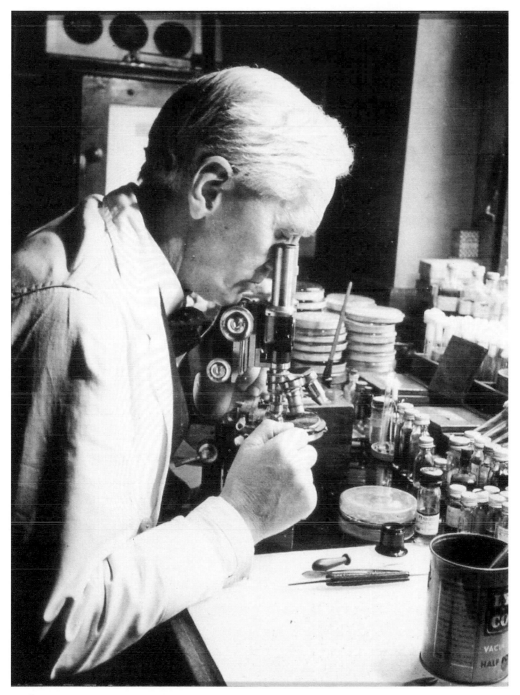

Although most doctors hated research, Fleming quickly developed a passion for studying the behavior of microscopic bacteria.

Soldiers in World War I faced a greater risk of dying from infection than from combat.

Chapter Five
The War With Infection

O n June 28, 1914 the Archduke Franz Ferdinand, heir to the Austrian throne, was assassinated in Sarajevo, Serbia. Austria, with the support of Germany, sent an ultimatum containing many demands to Serbia. The response was hostile, so Austria and Germany declared war on Serbia. But Russia was allied with Serbia, and Russian troops mobilized against the Austrian border. Other countries were quickly dragged into the conflict. Across the ocean, the United States and Canada would eventually be summoned into what became known as World War I.

For Alec Fleming, the war meant two things. One, he was immediately enlisted in the wartime medical effort and promoted to lieutenant. Second, he had an opportunity to test some of the theories that he and Almroth Wright had been formulating regarding infection.

Wright, who would be awarded the rank of lieutenant colonel, saw the conflict in personal terms as well. It had been during another war—the Boer War of 1900—when his advice regarding inoculations for typhoid was ignored. Within two months after World War I began, Wright published an article in the *London Times* called "On the Inoculation of Troops Against Typhoid Fever and Septic Infection."

By using figures from vaccinations in the US, Japan, Germany and other countries, Wright provided convincing arguments for the inoculation of British troops. This time his argument worked. Over the course of World War I, ten million doses of the anti-typhoid vaccination were prepared. As a result, the death rate from this infection was less than ten percent what it had been during the Boer War.

But Wright and Fleming's work at home was brief. In October of 1914, they were posted to a unit in Boulogne, France and began working at the British Army General Hospital, housed in a building known as "the Casino." Once elegant, the building was filled with beds so closely packed together it was difficult to move between them. Wounded soldiers, many still teenagers, arrived directly from the battlefield, where only the most basic of first aid methods had been applied. The men would still be in their dirty uniforms, their wounds barely treated.

Although the conditions endured by wounded soldiers in the makeshift hospital were bad, those suffered by Alec Fleming and the rest of Wright's men three floors below in the Casino's basement were even worse. Consisting of several dirty and poorly lit storerooms, the laboratory where Wright was expected to develop treatments for gas gangrene and other battlefield horrors was itself a breeding ground for infectious bacteria. Overhead pipes leaked raw sewage, and no amount of disinfectant covered the terrible stench.

Wright eventually was able to get his men moved to the top floor, an old fencing room. Well-lit and without leaky plumbing, the new location still had a problem. There was no water, gas or drainage.

Fleming used his skills as a craftsman. He constructed most of the Bunsen burners and gas blowers. He rigged up clean gas cans to supply water. Fleming would later describe the converted fencing room as the best lab he ever worked in.

Fleming's job was to identify the sources of the soldiers' infections. He took swabs of wounds from the living and cut into the tissue of the dead. He studied bullet fragments and bits of shredded shrapnel. He cut swatches from uniforms and slid them under a microscope lens.

Ninety percent of these samples grew *Clostridium welchii*, the bacteria which caused deadly gas gangrene. There were other dangerous bacteria as well, such as *Tetanus bacilli*. Both of these bacteria lived in horse intestines, surviving in the manure which littered the fields soldiers fought and died on during World War I.

The challenge was determining how organisms such as these which don't grow in the presence of oxygen—anaerobic organisms—could live in wounds which were exposed to oxygen.

In two papers Fleming wrote for *The Lancet* in 1915, he demonstrated how these wounds also were infected with organisms which flourish in oxygen, called aerobic organisms. These organisms would help the anaerobic organism survive. What was even more surprising was that the usual method of treating these men killed the white blood cells that their bodies needed to fight infection. Even worse, the technique allowed dangerous bacteria to thrive because they had been driven so far into the body by the force of the injuries.

The method of treatment Fleming was fighting involved the use of antiseptics. The word "septic," which comes from the Greek word for "rotten," was first used in a medical sense by Sir John Pringle in 1750. He'd noted how battle wounds became rotten and he suggested "antiseptics" to fight the rot.

It wasn't until the 1870s that Joseph Lister, among others, realized this "rot" was caused by microorganisms—bacteria. Because of this, surgeons became far more careful about germs, and began wearing sterile operating gowns, masks, and eventually gloves.

It was Lister's recommendation to use antiseptics—substances which would kill the germs but not living tissues—which would become the basis for Fleming's papers. Unfortunately Lister's antiseptic, which was highly diluted carbolic acid, proved ineffective in treating those injured in combat, because infectious bacteria were driven in too deeply to be affected.

If Fleming hadn't already learned about the challenges he would face when fighting established medical wisdom from his boss, the reactions to his opinions about antiseptics provided an effective lesson. Fleming believed that antiseptics dangerous to white blood cells should be banned. Instead the wound should be irrigated with a sterile salt solution. Dirt, dead tissue and any other bacteria-carrying matter were to be removed as much as possible.

Wright advocated this position in a widely circulated memo. He immediately ran into trouble. Sir Arthur Sloggett, a powerful Army doctor, claimed Wright had exceeded his authority and should be removed from his position. In the end, Wright's team was the only one that stopped using antiseptics. Despite the risks, the antiseptic method continued to be used by everyone else in the military.

Although Fleming's position on antiseptics was ignored, his work studying bacterial infections gained him respect. And his professional reputation wasn't the only thing changed during the war. On December 23, 1915, Alec Fleming married Sally McElroy, the owner of a successful nursing home. Because she was such an outgoing woman, many people believed she was the one who proposed.

After the wedding, Alec Fleming returned to the Casino. He told everyone about his new wife. No one believed him. Finally he displayed a photo taken at the ceremony. If it hadn't been for that, they would have assumed that shy Fleming was still a bachelor.

Marriage wasn't the only diversion for the doctor. Fleming enjoyed playing golf, indulging in his own peculiar techniques. Sometimes he played entire rounds using just his putter, sometimes sinking the ball using his golf club like a pool cue. He even played a practical joke on an arrogant staff colonel. Whenever the man hit a drive, Fleming would wait until he was out of eyesight and drop the ball in the hole. The colonel became convinced that he was regularly hitting a hole in one.

World War I finally ended in November, 1918. Two months later, Fleming returned home to St. Mary's. The doctor's first great discovery was right under his nose. Literally.

Like all of the doctors in his department, Fleming used himself as a guinea pig. Vaccines were regularly tested on the staff of Wright's inoculation department. The doctors who took part in these experiments often put their lives at risk. Indeed, two doctors died after being injected with a dangerous virus.

Even Alec's family wasn't safe from the practice. As his brother Robert recalled in his memoir, "I must have had my arm punctured and injected with hundreds of different kinds of dead microbes in those days. Alec must have jabbed killed microbes into himself many more times. Here again Alec was a master of technique. He could insert the needle of a syringe into you so that you hardly felt it."

This focus on discovery didn't take a rest when Alec came down with a bad cold in the fall of 1921. He was in his kitchen working over some petri dishes when a sneeze caught him by surprise and he sprayed mucus all over a dish. Instead of cleaning it, he decided to see its effect on the bacteria covering the dish's surface. He also swabbed more mucus on other bacteria-laden dishes.

Later examinations revealed the mucus was killing the bacteria. He realized this could be an enormous discovery. Intrigued, Fleming began judging other secretions.

When he recovered from his cold, he returned to his lab. He decided to see if tears had the same result. He used his own tears plus those of his colleagues and any assistant who happened to walk into his lab. V.D. Allison, a colleague of his, later wrote that "Many were the lemons we had to buy to produce all those tears! We used to cut a small piece of lemon peel and squeeze it into our eyes, looking into the mirror of the microscope."

A single tear dissolved a colony of bacteria in a few seconds. "I had never seen anything like it," Fleming later wrote.

The discovery was Fleming's first real chance to step out of Wright's substantial shadow. Fleming published papers dealing with the substance and spoke at lectures. Unfortunately his quiet, shy manner worked against him when he described his work. Few of his listeners appreciated it as any type of important breakthrough. But Fleming pressed on, studying both human and animal samples. He found the same germ-killing substance in many things. The richest source was egg whites, where it was one hundred times more concentrated than in tears.

Fleming convinced Wright to submit his paper to the Royal Society, the oldest scientific club in the world. Wright called the substance lysozyme, but his presentation before the distinguished scientists was poorly received. They regarded the discovery as just a curiosity. But the scientists' indifferent reaction didn't change Fleming's opinion. He believed his discovery could someday be used to successfully treat bacterial infections.

He worked with a variety of doctors and scientists, both at St. Mary's and elsewhere. "I realized that every living thing must, in all its parts, have an effective defense mechanism, otherwise no living organism could continue to exist. The bacteria would invade and destroy it," Fleming told a biographer. Believing he was on the brink of a breakthrough, Fleming would later write, "If we had this substance pure, it ought to be possible to maintain in the body a concentration which would kill bacteria."

Fleming's discovery of lysozyme would lead others to write thousands of papers about its potential. Despite that, it never evolved into the bacteria killer he had imagined. The skilled rifleman had infectious bacteria in his sights. He just needed a better weapon.

Called "Almost Wright" by his critics, Almroth Wright stood by his conviction that in the future, most doctors would be immunizers, and that most of his peers offered cures that didn't work.

Chapter Six

Eureka!

● ●

Now in his forties, Alec Fleming settled into a comfortable life. A few years before, he and his wife Sally had bought a second home in rural Burton Mills and named it "The Dhoon." Alec would enjoy spending weekends and other free time at The Dhoon for the next three decades. He would putter around the three-acre estate, working in the garden and returning to his childhood spirit, when nature played a huge role in his life.

In 1924, Sally gave birth to a son, Robert. Alec gave up his golf game to spend time with what would be his only child. He was a devoted, if somewhat absentminded father. A friend would later tell the story of a fishing outing when Alec was more concerned about reeling in a fish than with his son, who had accidentally fallen overboard. Robert was thankfully rescued unharmed.

Still, despite the quiet pleasures of his middle age, Alec suffered under a nagging sense that his time to make a breakthrough discovery was rapidly fading. Although he continued to promote the germ-killing potential of lysozyme, hardly anyone else was interested.

In 1926, Wright retired from St. Mary's at the age of 65 (although he would still advise the school). Fleming was promoted to the newly created Chair of Bacteriology, which meant he was a full professor in the pathology department. This gained him considerable power in selecting his experiments. Although Fleming enjoyed the advancement, he rarely used his title, partially because of his modest nature and partly out of respect for a peer, John Freeman, who had wanted the job.

Fleming was now older than Wright had been when he began running the pathology department. However, by that time Wright had already made a huge breakthrough with the anti-typhoid vaccine. Although Fleming continued to devote a great deal of time to his lab work, he made little progress.

Then it happened.

During an unusually warm London September in 1928, Fleming took a two-week vacation. Because of the hot weather, the windows of his lab were open and he forgot to close them when he left. In his haste, he didn't bother to clean up an old culture plate that he had smeared with staphylococcus bacteria, which was a notorious cause of skin infections.

Soon after he left, a spore containing a rare strain of a fungus mold called penicillium drifted into his laboratory from another lab in the same building. By chance, it settled onto that messy culture plate.

And as if that wasn't lucky enough, the weather stepped in to add even more. The temperature briefly dropped, so the mold began to grow. Then things heated up again, and the bacteria on the plate sprouted like a weed. Except in one spot.

That one spot attracted Fleming's eye when he came back from his vacation. It was where the penicillium spore had settled and grown, looking like blue fuzz on old cheese.

Fleming believed he had discovered something important, that the mold contained a substance that prevented bacteria from growing.

"Instead of casting out the contaminated culture with appropriate language," Fleming told a biographer, "I made some investigations."

So instead of cursing the filth, Fleming prepared a slide.

Beneath the microscope, he identified the brushlike hairs of the penicillium mold. He then scraped the culture plate into a nutrient broth. Over several days, he would observe the mold's change from white to green to black.

What was most interesting to Fleming was not the mold's change in color. It was the bright liquid which formed a halo beneath the mold. Inside the halo, the bacteria were dead. When Fleming tested the liquid on other bacteria, it killed them as well.

Others were not as impressed by Fleming's discovery. He showed it to several colleagues, including the school's resident mold expert, J.C. La Touche. "Old Moldy," as his peers and students called him, identified the mold more specifically as *Penicillium rubrum*. Later it would be proven to be not *rubrum*, but an even rarer variant known as *notatum*. Besides not identifying the mold correctly, Old Moldy was unimpressed with its effects. After all, the mold had been used on skin infections for over two thousand years. So the fact that it killed a particular bacteria was hardly surprising.

Fleming didn't see it that way. "There are thousands of different molds and there were thousands of different bacteria," Fleming wrote much later. "...that chance put the mold in the right spot at the right time was like winning the Irish Sweep."

Fleming believed the strain he'd identified was much more powerful than something ancient people used to rub on their face. Scientists, including Fleming, had long been developing artificial chemicals to fight germs. But his own experience in World War I was good evidence for how useless this was. Fleming believed he'd discovered an important antibiotic—a substance in nature that kills germs.

All he had to do was prove it.

He began by collecting other molds—molds that grew around Paddington, samples he got from other scientists, even the mold from the soles of old shoes. He tested all of them on bacteria. None was as effective a killer.

Fleming began breeding the mold he'd first discovered, placing it in agar, which is a gelatin-like substance derived from seaweed. He tested it on many varieties of bacteria. It killed them all. Even at 1/1000 strength, the mold continued to work.

Fleming worked with Stuart Craddock, who'd collaborated with Fleming on an earlier ineffective germ killer. He brought in a pair of other young assistants, and devoted his time to purifying the antibiotic. They discovered it grew best at 20 degrees Centigrade, and that the growth all but disappeared when the temperature rose seventeen degrees. Fleming added tiny amounts of acid to help it grow.

To the others who worked at St. Mary's, "Little Flem's" work was a tremendous nuisance. Both Craddock and

Fredrick Ripley, a chemist who'd helped Fleming isolate lysozyme, were huge men. Whenever they strode the narrow hallways, other doctors would find it hard to get around them. The two men made the labs seem more crowded than they were. Even worse, the work of purifying the mold created an awful stink.

Fleming didn't notice these issues. He set about testing his discovery. Craddock became the first guinea pig when he suffered an infected sinus. A swab dabbed on the infected area was inconclusive. Even less encouraging was the test on a dying amputee. The patient was unresponsive to the new antibiotic and died.

The greatest problem was that the new antibiotic, which Fleming now called penicillin, was filled with impurities that reduced its effect. Ridley worked to remove these by heating the mixture, until it was fifty times stronger than the broth.

Craddock later wrote, "We were full of hope when we started, but as we went on, week after week, we could get nothing but this glutinous mass which...would not keep."

On February 13, 1929 Fleming gave his first lecture on his discovery to the Medical Research Club. He spoke the way he always did: quietly and haltingly. He might have been sure in his head and his heart, but it never showed in his voice.

When Alec Fleming concluded the lecture he waited for questions. And waited. Not a single hand was raised, not a single question was asked. No one in the room seemed to think Alec's second breakthrough would get any further than his first.

World War II spurred the production of penicillin. Though it was discovered in England, it was manufactured in places like New Brunswick, New Jersey, because of Nazi attacks on London. Fleming often examined the manufacturing of his penicillin, as he did here with a Squibb employee in 1945.

Chapter Seven

Another Try

• •

For the second time in his career, Alec Fleming had been convinced he was onto a major scientific breakthrough, only to have other scientists ignore it. The pair of discoveries were happy coincidences: lysozyme the product of a sneeze, penicillin the product of a hot, humid lab and unwashed plates. Perhaps the coincidence of his discoveries eased the pain a bit, for by the summer of 1929 his work on the mold diminished.

Although Fleming would occasionally try new ways of purifying penicillin, and would write a few pages of notes here and there, he had turned to other areas of science within a few years. Fleming contributed to books, spoke at lectures and enjoyed traveling with his wife and young son. He also responded to requests from various schools for samples from his mold. One of those schools was Oxford.

Located fifty miles away from London, Oxford is one of the most prestigious universities in the world. In the late 1930s, a pair of researchers would follow the path Fleming had begun to clear. As is often the case, penicillin needed others to take it from its original discovery to the point where it could be successfully utilized.

Howard Florey was a 37-year-old professor of pathology who had read Fleming's papers for years. Earlier he'd overseen the extraction of semi-pure lysozyme for the first time, and was fascinated by what that germ killer might someday accomplish. So even as Wright was scoffing at Fleming's discovery, and Fleming was unable to get funding for penicillin research, Florey was guiding his department towards a breakthrough.

Ernst B. Chain was a German Jew who'd fled Berlin when Adolf Hitler's rise to power threatened his very life. Chain was a chemist; at age 29 he was made head of Oxford's biochemical department, supervised by Florey. In the beginning his research was focused on lysozyme and he read everything Fleming wrote.

Penicillin itself was an accident. So too was Chain's interest in it. In 1938, the young scientist discovered Fleming's paper on the mold, now nearly a decade old. "It was sheer luck that I came across Fleming's paper," he was quoted in a biography of the doctor. "No chemist would normally think of reading a work on pathology to assist his research in chemistry."

The paper sparked Chain's interest and he decided to take up where Fleming had left off. As one more coincidence, Chain quickly learned that Oxford had cultures of Fleming's original mold in its lab.

Florey shared Chain's enthusiasm and applied for a grant from the United States-based Rockefeller Foundation. Oxford received $5,000, and the work began.

As the pair of researchers devoted their time to reducing the impurities in penicillin, fifty miles away the mold's discoverer was assisting with another battle. Adolf Hitler was leading Germany as the country invaded and attacked countries across Europe. On September 3, 1939, Britain declared war after Germany invaded Poland. Everyone believed the front lines of World War II would no longer be across the English Channel in France. The

Germans had the air power to attack London directly. Indeed, during the course of the war, Fleming's London home would be shook by several bombing attacks.

Fleming volunteered quickly and was put in charge of supervising the efforts of the Ministry of Health. He realized he was facing another horrific war, one where infectious bacteria would again be many soldiers' deadliest enemy.

Florey and Chain were working hard to change this. After Florey discovered that penicillin's instability was heightened by evaporation—the method Fleming and his assistants had relied on—Chain turned to another technique. He utilized a freeze-drying process that wasn't available when Fleming was doing his work. Called lyophilization, it enabled liquid penicillin to move from liquid to solid while in a vacuum state. The brownish yellow powder which resulted proved to be purer than anything Fleming had found. To make it even more pure, Chain added it to methanol, which reduced the protein and salt impurities.

They tested groups of mice infected with streptococcus germs. Out of twenty-five injected with the penicillin, all but one recovered. Those left untreated, died.

The penicillin wasn't perfect. It was, however, one thousand times stronger than Fleming's concoction and ten times more effective than the strongest sulfonamide, the current treatment for infections.

In August of 1940, the pair published their findings in *The Lancet*. Their article had at least one very interested reader: Alec Fleming! He had no idea anyone was working on his discovery. When Fleming called up Florey at Oxford, the younger doctor was stunned. He'd thought Fleming had been dead for years.

World War II contributed immensely to the rapid deployment of the new drug. Thousands of wounded soldiers desperately needed something to protect them from death by infection. With bombs falling on England, it was safer to do the actual manufacturing in the United States.

So Florey traveled to Peoria, Illinois, because an agricultural research center there had developed a successful technique for fermentation, which was essential to grow penicillin. Scientists quickly designed and built large "deep fermentation" tanks which used a steady stream of sterilized air to produce penicillin.

Even better, Illinois had lots of corn, a vegetable unknown in England. Corn steep liquor, a byproduct of corn processing that had been considered as a waste product, supplied nutrients that helped to substantially increase the yield.

People who were involved in producing penicillin were encouraged to bring in rotting fruits and vegetables. A researcher named Mary Hunt found a cantaloupe in a supermarket with a golden mold that turned out to be a strain called *Penicillium chrysogenum*. It grew so well that it nearly doubled the total penicillin output.

All these discoveries made it possible to produce penicillin commercially, and it saved the lives of countless thousands of soldiers, sailors and airmen during World War II.

Back in England, Fleming was knighted for his discovery by King George VI in 1944. Because of a concern about bombing attacks, the honor took place in the basement of Buckingham Palace. The next year, he shared the highest honor a scientist can receive when he was awarded the Nobel Prize for Medicine along with Chain and Florey. Although the Oxford scientists received their share of recognition, it was Fleming who was treated like a hero.

Unlike many other scientists in the 20th century, Fleming enjoyed his newfound fame. He appreciated being recognized, despite his shyness. He traveled the world and spoke widely about his discovery. Still, he never completely lost his quiet delivery. But with all his honors and fame, this time people listened to every word he said.

On October 28, 1949, Alec's wife Sally died. Fleming suffered greatly. He'd hoped to spend the rest of his life with the outgoing woman, who in later years insisted that everyone call her "Sareen" because Sally was too boring a name for her.

"My life is broken," Fleming told a friend.

For several years, Fleming would work and study alone. But in April of 1953, he married Amalia Voureka, a much younger bacteriologist. The two shared the rest of his life together. On March 11, 1955, Fleming died of a heart attack. His death was mourned around the world.

By then, pharmaceutical corporations had begun manufacturing synthetic penicillin, which created a huge new industry. Because of Fleming's discovery, diseases like gas gangrene, syphilis, and tuberculosis have almost completely disappeared in industrialized countries.

None of this would have happened if Fleming had been discouraged by how little other people thought about his discovery. Yet he persevered, later saying in a magazine interview that, "My only merit is that I did not neglect the observation and that I pursued the subject as a bacteriologist."

Because Fleming managed to be both persistent and messy, the lives of countless millions were saved. He discovered the most effective life-saving drug in the world, and literally changed the course of history.

Fleming (third from left) was well into middle age before his discovery of penicillin was appreciated. Here he receives a humanitarian award, along with a silver plaque and $1,000, by the Varsity Clubs of America for "unusual and unselfish service to humanity."

Alexander Fleming Chronology

- 1881, born in Ayrshire, Scotland.
- 1895, moves to London, England and lives with brother Tom.
- 1897, leaves Polytechnic Institute due to lack of money, takes job as a shipping clerk.
- 1901, enrolls in St. Mary's Hospital Medical School.
- 1905, passes primary exam for the Fellowship of the Royal College of Surgeons.
- 1906, after taking final exams, joins Almroth Wright's inoculation department at St. Mary's.
- 1908, publishes first scientific paper in *The Practitioner*.
- 1909, passes exams qualifying him to be a surgeon, but never practices surgery again
- 1915, marries Sally McElroy.
- 1921, discovers lysozyme, a bacteria-fighting enzyme produced by the body
- 1924, his only child, Robert, is born.
- 1928, discovers penicillin and its antibiotic properties.
- 1940, working from Fleming's original culture and medical papers, Oxford researchers Howard Florey and Ernst B. Chain successfully test a purer, stronger version of penicillin.
- 1944, is knighted by King George VI.
- 1945, receives Nobel prize for Medicine.
- 1949, wife Sally (Sareen) dies.
- 1953, marries Amelia Voureka.
- 1955, dies of a heart attack.

Penicillin Timeline

- **500 BC:** First documented use of molds to fight skin infections.
- **1780s:** English doctor Edward Jenner shows that injecting weakened bacteria can prevent disease.
- **1865:** Louis Pasteur relates theory to Joseph Lister that specific diseases are caused by specific bacteria.
- **1870:** dramatic decline in post-operative infection during surgery in the Franco-Prussian war is credited to Lister's call for clean operating rooms and the use of diluted carbolic acid.
- **1870s/80s:** German doctor Robert Koch's work leads to creation of bacteriology, the study of bacteria.
- **1875:** John Tyndall describes anti-bacterial properties of penicillin.
- **1882:** Russian-French bacteriologist Elie Metchnikoff observes white blood cells' ability to fight infection.
- **1915:** Fleming begins to disprove some of Lister's theories about antiseptics.
- **1921:** Fleming discovers lysozyme.
- **1928:** Fleming discovers mold penicillium's antibiotic properties.
- **1940:** Howard Florey and Ernst B. Chain succeed in removing impurities from penicillin and strengthening it.
- **1941:** first human tested with penicillin.
- **1945:** penicillin made available to general public following World War II's conclusion.

Further Reading

Books

Hughes, W. Howard, *Alexander Fleming and Penicillin*. New York: Crane-Russak, 1977.

Otfinoski, Steve. *Alexander Fleming: Conquering Disease with Penicillin*. New York: Facts on File, 1992.

Bullock, W.A.C. *The Man Who Discovered Penicillin: The Life of Sir Alexander Fleming*. London: Faber and Faber, 1963.

Web Sites

www.Nobel.se/Medicine

www.SJSU.edu

www.Amaz.com/nobel

www.biography.com

www.time.com/time/time100/scientist/profile/fleming.html

Glossary of Terms

Antibiotic - a natural substance which destroys germs

Antiseptic - a chemical substance which destroys germs

Bacteria - microscopic single-celled organisms; some are dangerous, some are helpful, most are harmless

Bacteriologist - scientist who studies bacteria and looks for ways to fight diseases they cause

Broth - a nutritive soup used to grow organisms

Culture - a growth of bacteria used for scientific study

Infection - a disease caused by germs

Mold - a fungus growth

Pathology - the study of disease and its causes

Penicillin - antibiotic produced by molds; first discovered by Fleming

Penicillium - a category of fungus

Petri dish - a shallow, round glass or plate used to prepare bacteria cultures

Strain - a particular variety or group of microbes

Index